A tiny piece to a bigger puzzle.

Published by the Honey Bee Company LLC
2906 Central Street • Suite 258
Evanston, IL 60201
847.864.8032

ISBN 978-1-947933-01-9

MY NAME IS...

I'M READY TO GET TO WORK!

Scan this QR code to read the Paige & Paxton Discover a Small World storybook before you begin!

Paxton has a sore throat.

His mom has taken him to the doctor.
His sister, Paige, has come too.

Color the pictures.

Have you ever been to a doctor?

Draw a picture of your doctor here.

Circle the things you might see in a doctor's office.

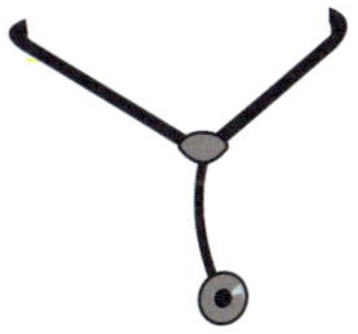

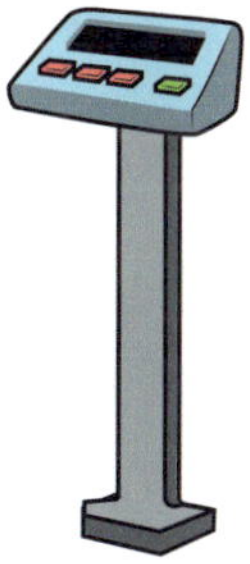

How many things did you circle? ____________

Paxton's doctor thinks tiny, tiny creatures might be making Paxton sick.

Oh no!

Tiny, tiny creatures called

Microorganisms

Microorganisms are creatures too small to see with our own eyes.

If we can't see them, how do we know they're here?

Circle the largest microorganism.

Sometimes we can see them when they're standing in a crowd.

Can you see microorganisms in the petri dishes?
Circle the petri dish with the most microorganisms.

We can see them with a microscope.

Count by twos to connect the dots.

A microscope is a powerful tool that makes things appear bigger than they are.

How many microscopes do you see?

There are __________ microscopes.

We want to know all about microorganisms because they're a big part of our world.

Microorganisms are everywhere!

Circle the places you might find microorganisms.

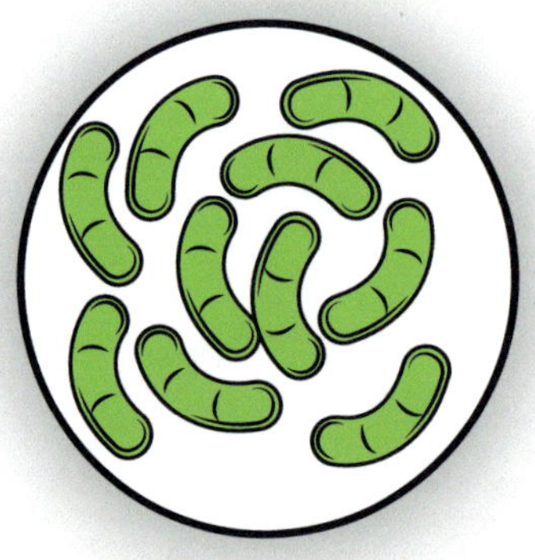

Microbiology

is the way we find out about microorganisms.

Microbiology is the way we find out how these tiny, tiny creatures work and why they are the way they are.

Microbiology is science.

Microbiologists

are scientists who try to find out how microorganisms work and why they are the way they are.

Paige and Paxton want to know all about microorganisms.

Help Paige & Paxton get to the microbiologist.

The microbiologist tells Paige and Paxton all about microorganisms.

Draw lines to the microorganisms that are the same.

Paige and Paxton learn about the good things that they do.

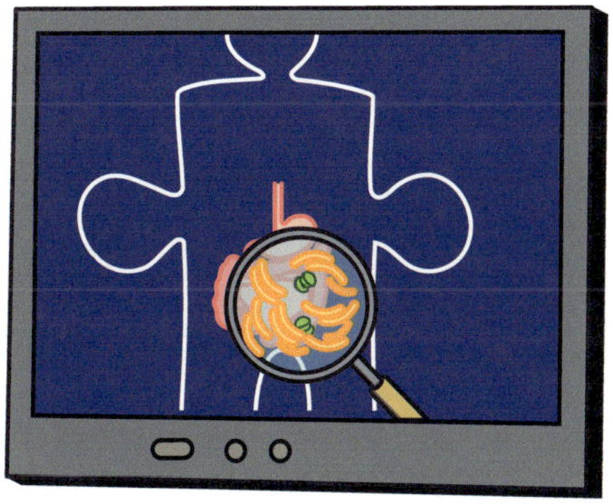

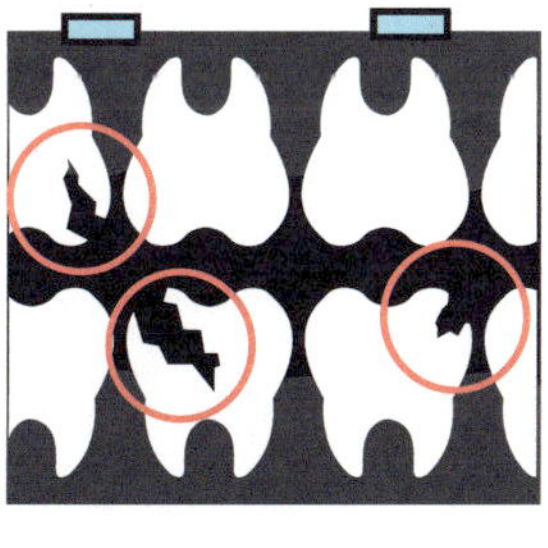

Circle the good things that microorganisms do.

Paige and Paxton learn about the bad things that they do.

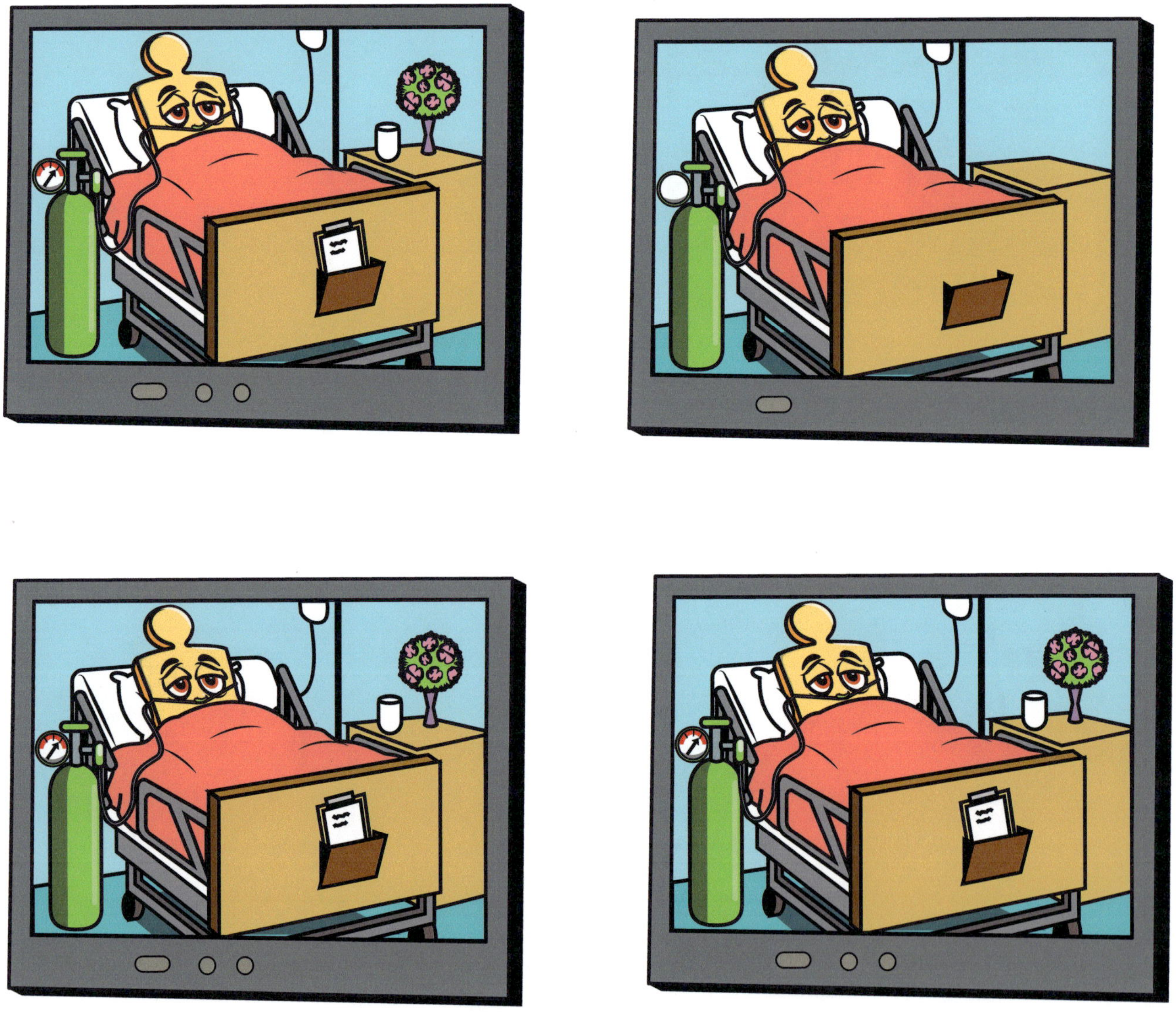

Circle the picture that's different.

They learn that microorganisms are just like us. There are good ones and there are bad ones!

Draw lines to the ones that are the same.

Now Paige and Paxton know all about microorganisms. And so do you!

Circle the smallest picture.

Who would have thought that learning about microorganisms would be the best piece of Paxton's day? And yours too!

Draw a picture of something you've learned about microorganisms.

OUR MISSION AT PAIGE & PAXTON LABS IS TO MAKE STEM (SCIENCE, TECHNOLOGY, ENGINEERING, AND MATH) EDUCATION EASY AND FUN FOR CHILDREN, STARTING FROM A YOUNG AGE.

We believe that it's never too early to start learning and developing skills in these areas and hope that by making it easy for parents and educators, we can change the trajectory of an entire generation.

Paige & Paxton Labs is a revolutionary system of STEM education and parenting support created by best-selling authors, award-winning educators, and accomplished STEM professionals. Designed to introduce STEM concepts to children as young as three years old, Paige & Paxton Labs includes:

- A series of illustrated STEM books and workbooks that both captivate and teach
- Curriculum for kids to explore STEM concepts through imaginative storytelling and active learning projects
- A growing network of parents who want to share their love of science with their children

JOIN THE COMMUNITY:

Like Us on Facebook

Pin Us On Pinterest

Follow Us On Twitter

Made in the USA
Monee, IL
05 November 2026

67380372R00019